International Environmental Labelling

Vol.3 of 11

For All People who wish to take care of Climate Change
Fashion & Textile Industries:
(Fashion Design, The Fashion System, Fashion Retailing,
Marketing and Marchandizing, Textile Design and Production,
Clothing and Textile Recycling)

Jahangir Asadi

Vancouver, BC CANADA

Suggest an ecolabel

If you think that we missed a label and/or you are an ecolabelling body, please consider to submit for the next editions of our 11 Volumes International Eco-labelling Book series. Please send your details, and we'll review your suggestions. Our goal is to be as comprehensive as possible, so thank you for your help!
info@TopTenAward.Net

Published by: Top Ten Award International Network
Vancouver, BC **CANADA**
Email: Info@TopTenAward.net
www.TopTenAward.net

Ordering Information:
Quantity sales. Special discounts are available on quantity purchases by universities, schools, corporations, associations, and others. For details, contact the "Sales Department" at the above mentioned email address.

International Environmental Labelling Vol.3/J.Asadi—2nd ed.
ISBN:978-1-7773356-5-6

Contents

I dedicate this book to my mother in law: Setareh.

We hope that, 10,000 years from now, future generations will be able to see flowers that provide bees with nectar and pollen and... BEES provide flowers with the means to reproduce by spreading pollen from flower to flower...

Jahangir Asadi

Acknowledgements:

I wish to thank my committee members, who were more than generous with their expertise and precious time. I would like to acknowledge and thank the Top Ten Award International Network for allowing me to conduct my research and providing any assistance requested.

It should be noted that all the required permissions for using the logos and trade marks has been obtained to be published in this volume.

Do you know that:

"Eco-friendly" fibres

means that their production process has a low impact on the environment and mets at least half of the below criteria:

- Low water needs
- Low Energy need
- Chemicals control
- Biodegradable
- From renewable resource
- Made of wastes

Top Ten Award International Network

Top Ten Award international Network (TTAIN) was established in 2012 to recognize outstanding individuals, groups, companies, organizations representing the best in the public works profession.

TTAIN publishing books related to international Eco-labeling plans to increase public knowledge in purchasing based on the environmental impacts of products.

Top Ten Award International Network provides A to Z book publishing services and distribution to over 39,000 booksellers worldwide, including Apple, Amazon, Barnes & Noble, Indigo, Google Play Books, and many more.

Our services including: editing, design, distribution, marketing

TTAIN Book publishing are in the following categories:

Student
Standard
Business
Professional
Honorary

We focus on quality, environmental & food safety management systems , as well as environmnetal sustain for future kids. TTAIN also provide complete consulting services for QMS, EMS, FSMS, HACCP and Ecolabeling based on international standards.

ISO 14024 establishes the principles and procedures for developing Type I environmental labelling programmes, including the selection of product categories, product environmental criteria and product function characteristics, and for assessing and demonstrating compliance. ISO 14024 also establishes the certification procedures for awarding the label.

TTAIN has enough experiences to help create new ecolabeling programmes in different countries all over the world.
For more detail visit our website : http://toptenaward.net
and/or send your enquiery to the following email:
info@toptenaward.net

CHAPTER 1

Introduction

This book is dedicated to the subject of environmental labels. The basis for the classification of its parts goes back to the types of environmental labelling according to the classifications provided by the International Organization for Standardization. In each section, while presenting the relevant definitions, I mention the existing international standards and present examples related to each type of labelling. Environmental labelling is an important and significant topic, and its richness is added to every day, which has attracted the attention of many experts and researchers around the world. The idea of compiling this book, came to my mind when I observed that national environmental labelling models have been developed in most countries of the world, but in many other countries, the initial steps have not been taken yet. Therefore, I decided to create the first spark for the development of environmental labelling patterns in other countries by collecting appropriate materials and inserting samples of labelling patterns of different countries of the world. It should be noted that the description of each environmental label in this book does not indicate their approval or denial; they are included only to increase the awareness of all enthusiasts and consumers of the meanings and concepts derived from such labels. We hereby ask all interested parties around the world who wish to start an environmental labelling program in their country to benefit from our intellectual assistance and support in the form of consulting contracts. Increasing human awareness of the urgent need to protect the environment has led to changes in all levels of activities, including the production of marketing products, consumption, use, and sale of goods and services at the national and international levels. Stakeholders involved in environmental protection include consumers, producers, traders, scientific and technological institutes, national authorities, local and international organizations, environmental gatherings, and human society in general. Decisions by consumers and sellers of products are made not only on the basis of key points such as quality, price, and availability of

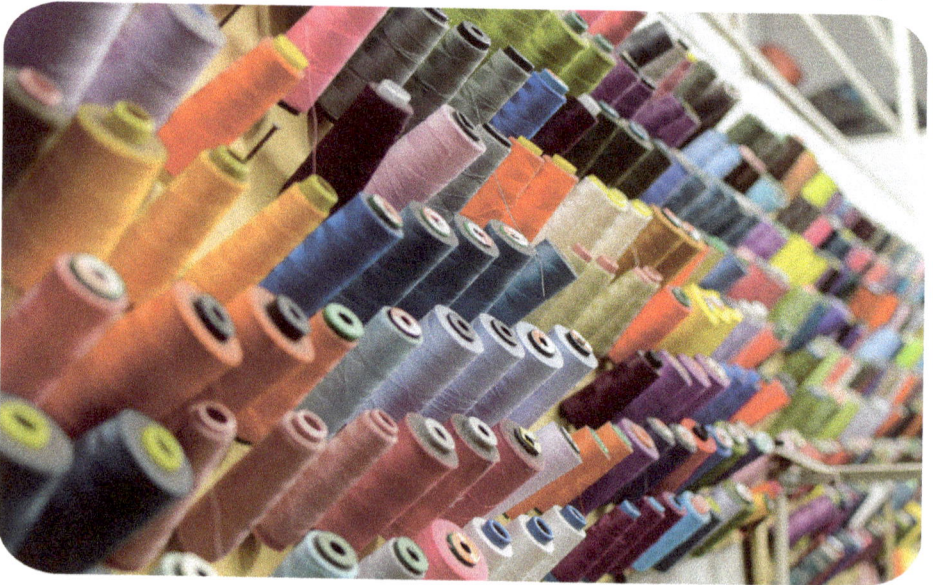

products but also on the environmental consequences of products, including the consequences that a product can have before, after and during production. The most important environmental consequences include water, soil, and air pollution along with waste generation, especially hazardous waste. Further consequences include noise, odor, dust, vibration, and heat dissipation as well as energy consumption using water, land, fuel, wood, and other natural resources. There are further effects on certain parts of the ecosystem and the environment. In addition, the environmental consequences not only include the natural use of the products but also abnormal and even emergency or accidental uses. The basis of studies and studies in this field is done through product life cycle evaluation, which generally involves the study and evaluation of environmental aspects and consequences of a category (product, service, etc.) because of the preparation of raw materials for production until they are used or discarded. Sometimes the phrase "review from cradle to grave" is used for such an evaluation. In addition to the above, the environmental consequences that may occur at any stage of the product life cycle, including the preliminary stages and its preparation, production, distribution, operation, and sale, should also be considered when evaluating it. This type of evaluation refers to product life cycle analysis from an environmental point of view,"

which is a useful tool for measuring the degree of environmental health of a product, comparing different products, improving product quality, and confirming the environmental health claims of the product. The environmental health analysis tool for products and services facilitates their placement in domestic or foreign markets, considering that the awareness of consumers and retailers about the environmental consequences of the product has increased, as has the accurate and explicit measurement by the people in charge at all levels. Local, national, and international in the field of environmental protection. Products that can claim to be environmentally complete in all stages of their life cycle and meet the mandatory and optional environmental needs are considered successful products. Environmental messages refer to the policies, goals, and skills of product manufacturing companies as part of the environmental management systems in which they are applied, and consumers and retailers are increasingly paying attention to this issue when making purchasing decisions. In addition, companies have been encouraged and even forced to adapt their environmental management systems to agencies and retailers and to local, national, international, and other environmental issues.

The environmental health message of a product can be conveyed to the consumer in various ways, including implicitly or explicitly. For example, the implicit or implicit message conveyed directly by the product to the customer is that the product is suitable for the intended use and purpose, and, without material waste in size, weight, and dimensions, is perfectly proportioned and without additional packaging. Sometimes it is necessary to convey these messages and claims about the correctness of the product quite clearly through magazines or other media as well as through certificates that are accurate, simple, and convincing to the consumer in the form of a label. These messages must be accurate and fact-based; otherwise they will nullify the product and create contradictory effects. Confirmation of these claims by a third-party organization will increase its credibility. It should also be noted that the multiplicity of these messages, depending on the type of products or companies producing them, confuses consumers in the market and also creates artificial boundaries or causes a differentiated distinction against certain products or companies. Various models, principles, and methods have been provided by local, regional, national, and international organizations to demonstrate product life cycle analysis and other guidelines on environmental management systems and their labels. At the national level, significant advances have been made in the design of environmental labels in various countries, including developing countries and the Scandinavian countries. For example, the first project was designated in Germany as a Blue Angel in 1977, later on Canada in 1988, the Scandinavian countries and Japan in 1989, the United States and New Zealand in 1990, India, Austria, and Australia in 1991, And in 1992, Singapore, the Republic of Korea, and the Netherlands developed their national environmental labelling. Environmental labels are an environmental management tool that is the subject of a series of ISO 14000 standards. These environmental labels provide information about a product or commodity in terms of its broad environmental characteristics, whether it is about a specific environmental issue or about other characteristics and topics.Interested and pro-environmental buyers can use this information when choosing products or goods. Product makers with these environmental labels hope to influence people's purchasing decisions. If these environmental labels have this effect, the share of the product in question can increase, and other suppliers may create healthy environmental competition by improving the environmental aspects of their products and commodities. The overall goal of environmental labels is to convey acceptable and accurate information that is in no way misleading regarding the environmental aspects of products and commodities, and they encourage the consumer to buy and produce products that reduce stress on the environment. Environmental labelling must follow the general principles that the International Organization for Standardization has published in a collection entitled the ISO 14020 standard, which refers to these general principles here. It should be noted that other documents and laws in this field are considered if they are in accordance with the principles set out in ISO 14020.

Recycled polyester, often called rPet, is made from recycled plastic bottles. It is a great way to divert plastic from our landfills. The production of recycled polyester requires far fewer resources than that of new fibres and generates fewer CO_2 emissions.

CHAPTER 2

General Principles on Environmental Labelling

1 The First Principle: Evironmental notices and labels must be accurate, verifiable, relevant, and in no way misleading and/or deceptive.

2 The Second Principle: Procedures and requirements for environmental labels will not be ready for selection unless they are implemented by affecting or eliminating unnecessary barriers to international trade.

3 The Third Principle: Environmental notices and labels will be based on scientific analysis that is sufficiently broad and comprehensive, and to support this claim, the product must be reliable and reproducible.

4 The Fourth Principle: The process, methodology, and any criteria required to support the announcements on environmental labels will be available upon request all interested groups.

5 The Fifth Principle: Development and improvement of environmental notices and labels should be considered in all aspects related to the service life of the product.

6 The Sixth Principle: Announcements on environmental labels will not prevent initiative and innovation but will be important in maintaining environmental implementation.

7 The Seventh Principle: Any enforcement request or information requirement related to environmental notices and labels should be limited to the necessary information to establish compliance with an acceptable standard and based on the notification standards and environmental labels.

8 The Eighth Principle: The process of improving the announcement and environmental labels should be done by an open solution with interested groups. Reasonable impressions must be made to reach a consensus through this process.

9 The Ninth Principle: Information on the environmental aspects of the product and goods related to an advertisement and environmental label will be prepared for buyers and interested buyers from a group consisting of an advertisement and an environmental label.

Activities that will teach kids about sustainability:
Recycle, Pick up trash, Sort the garbage, Finding Ecolabels on products during shopping
Thinking about creative ways to upcycle old T-shirts

CHAPTER 3

Types of Environmental Labelling

At present, according to the classification provided by the International Organization for Standardization, there are three types of environmental labelling
patterns:

1 Type I labelling: This labelling is known as eco-labelling, and because it
is difficult to translate this word into many languages, it presents another
reason to adhere to a numerical classification system. In the content of
Type I labelling, a set of social commitments that creates criteria according to
the scientific principles on the basis of which a product is environmentally preferable is discussed. Consumers are then instructed in assessing environmental
claims and must decide which packaging is more important.

2 Type II labelling: refers to the claims made on product labels in connection
with business centers. This includes familiar claims such as recyclable,
ozone-friendly, 60% phosphate-free, and the like. This type of labelling
can be in the form of a mark or sentence on the product packaging. Some of them
are valid environmental claims—and some can be completely misleading. Usually, all countries have laws against deceptive advertisements, so why has the
International Organization for Standardization discussed this issue? The answer
is that it is not clear whether the environmental claims have a technical basis or
whether the ad is meaningless.

3 Type III labelling: is a distinct form of third-party environmental labelling pattern designed to avoid the difficulties that can result from type-one labelling. Technical committee for Environment of International organization for Standardization has undertaken a new project to standardize guidelines and Type III labelling methods. One of the main objections raised by industries to Type I labelling is the basis for its management.

1 DESIGN

2 CUTTING

TAILORING 4

5 IRONING

2

3

DRESS PATTERN

3

5

6

CLOTHING STORE

CHAPTER 4

Type I Environmental Labelling

Type I labelling: This labelling is known as eco-labelling, and because it is difficult to translate this word into many languages, it presents another reason to adhere to a numerical classification system. In the content of Type I labelling, a set of social commitments that creates criteria according to the scientific principles on the basis of which a product is environmentally preferable is discussed. Consumers are then instructed in assessing environmental claims and must decide which packaging is more important.

Type I adhesive has the following specifications:

A. Has an optional third-party template.

B. When the product meets a certain standard, the labelling of this product is included.

C. The purpose of this program is to identify and promote products that play a pioneering role in terms of environment, which means its criteria are at a higher level than the average environmental performance.

D. Acceptance/rejection criteria are determined for each group of products and are publicly available.

E. The criteria are adjusted after considering the environmental consequences of the product life cycle.

Examples of Type I Labelling:

In this section, and considering the importance of this type of labelling, I provide a description of some examples of Type I labelling related to some countries along with a   list of products on which this mark is placed.

Global

Mission:

Our mission is the development, implementation, verification, protection and promotion of the Global Organic Textile Standard (GOTS). This standard stipulates requirements throughout the supply chain for both ecology and labor conditions in textile and apparel manufacturing using organically produced raw materials. Organic production is based on a system of farming that maintains and replenishes soil fertility without the use of toxic, persistent pesticides and fertilizers. In addition, organic production relies on adequate animal husbandry and excludes genetic modification.

The fastest way to learn about GOTS is to watch our four minute Simple Show Clip: https://global-standard.org/resource-library/clips.

Contact detail:

Web: www.global-standard.org
Email: mail@global-standard.org

China

China Environmental United Certification Center (CEC), approved by the Ministry of Ecology and Environment of the People's Republic of China (MEE) and accredited by Certification and Accreditation Administration Committee of PRC, is a comprehensive certification and service institution leading in environmental protection, energy saving and low carbon areas. . CEC is committed to serve building national ecological civilization; and has carried out research on environmental protection, energy saving, low carbon development strategies and solutions; has been continuously improving and innovating green industry evaluation system on industrial green development and transition CEC is building a bridge between green production and green consumption by offering independent, impartial and high-quality evaluation and certification service for government, enterprises and the public. CEC is a state-owned, non-profit, legal entity of independent third-party certification. It integrates the certification resource from the former National Accreditation Center for Environmental Conformity Assessment, the Secretariat of China Environmental Labelling Products Certification Committee, Environmental Development Center of MEE, the Chinese Research Academy of Environmental Sciences and other institutions. Business areas includes: products certification, management systems certification, services certification, addressing climate change, energy-saving and energy efficiency certification, green supply chain assessment, environmental stewardship, green credit assessment and green manufacturing system evaluation. CEC also carries out standard establishment and research project and international cooperation and exchanges, etc.

Contact:
Website: http://en.mepcec.com/
E-mail: zhangxiaoh@mepcec.com , zhangxiaoh@mepcec.com

Korea Eco-Label

Republic of Korea

The Korea Eco-labelling is a certification system enforced by the Ministry of Environment and KEITI(Korea Environmental Industry & Technology Institute). Since its foundation in April 1992, the system has certified a wide range of eco-friendly products, which were selected as excellent not only in terms of their environmental-friendliness, but also for their quality and performance during their life cycle. Korea Eco-labelling is voluntary certification scheme to attach logo to products with superior environmental quality throughout their lifecycle to other products of the same use, and thus to provide product information to consumers. For 30 years, the scheme has launched plenty of eco-labelling product standards covering personal and household goods, construction materials, office equipment furniture, etc. It products categories which cover all aspects of products, such as reduction of use of harmful substances, energy saving, resource saving, etc. As of April 30th 2021, 169 criterias(=standards), and certifications for 18,250 products(4,549 companies) have maintained.

Contact:
Korea Environmental Industry & Technology Institute(KEITI)
Office of Korea Eco-Label Innovation
Address: 215, Jinheung-ro, Eunpyeong-gu, Seoul, Repulic of Korea
T: +82 2 2284 1518
F: +82 2 2284 1526
E: accolly@keiti.re.kr
W: www.keiti.re.kr

Hong Kong

The Green Council is a non-profit, tax-exempt charitable environmental stewardship organisation and certification body (Reg. No.: HKCAS-027) of Hong Kong established in 2000. A group of individuals from different sectors of industry and academics shared the vision to help build Hong Kong into a world-class green city for the future. They formed the Green Council with the aim of encouraging the commercial and industrial sectors to include environmental protection in their management and production processes. The Green Council is a non-profit, tax-exempt charitable environmental stewardship organisation and certification body (Reg. No.: HKCAS-027) of Hong Kong established in 2000. A group of individuals from different sectors of industry and academics shared the vision to help build Hong Kong into a world-class green city for the future. They formed the Green Council with the aim of encouraging the commercial and industrial sectors to include environmental protection in their management and production processes. The Green Council is a non-profit, tax-exempt charitable environmental stewardship organisation and certification body (Reg. No.: HKCAS-027) of Hong Kong established in 2000. A group of individuals from different sectors of industry and academics shared the vision to help build Hong Kong into a world-class green city for the future. They formed the Green Council with the aim of encouraging the commercial and industrial sectors to include environmental protection in their management and production processes.

Contact:
Website: https://www.greencouncil.org/hkgls
Email: info@greencouncil.org
Telephone: (852) 2810 1122

Sri Lanka

National Cleaner Production Centre (NCPC), Sri Lanka was set up by UNIDO in 2002, as a project under the Ministry of Industry to provide the technical expertise and support to the industry and business enterprises in order to prevent pollution and conserve resources by the application of Cleaner Production (CP) and other proactive environmental management tools. NCPC Sri Lanka is registered as a Company by Guarantee not for profit organization under the Act No. 7 of 2007. Over the past two decades, it has evolved as the foremost sustainability solution provider in the country.

The ISO 9001:2015 certified Centre is a registered Energy Service Company (ESCO) under Sustainable Energy Authority (SEA) and a registered consultant under Central Environmental Authority (CEA). It is a founding member of UNIDO/UNEP Resource Efficient and Cleaner Production Network (RECP Net), a global family of 52 NCPCs. NCPC Sri Lanka is a member of Climate Technology Centre & Network (CTCN) and associate member of Global Eco-labelling Network (GEN). Accordingly, we at National Cleaner Production Centre (NCPC), Sri Lanka has developed Eco Labelling scheme under the ISO 14024:2018 - Environmental labels and declarations. NCPC Eco labelling scheme developed, with the Support of United Nations Environment Programme, Under One Planet Network Consumer Information Programme for Sustainable Consumption and Production (CI-SCP).

Contact:
Tel: +94 11 2822272/3,
Fax: +94 11 2822274
E mail: info@ncpcsrilanka.org
Web: www.ncpcsrilanka.org

Taiwan

The Green Mark GM) Program was launched by the Environmental Protection Administration of Taiwan (TEPA) in 1992. As the official Type I eco-labeling program, it is in compliance with the requirements of the international stadard, ISO 14024 and is considered an important tool to promote green consumption and production .

To improve the GM application/review mechanism and introduce a third party certification scheme, TEPA promulgated the «Guideline for the Management of Certification Organizations for Environmental Protection Products" in June 2012. Both Environment and Development Foundation (EDF) and the Taiwan Testing and Certification Center (ETC) were commissioned by TEPA as official certifiers. With the expansion of certification capacity and authorization of the certification decision, the certification time was greatly reduced.

Contact :

Website: www.edf.org.tw
TEL: 886-3-5910008 #39
E-mail: lhliu@edf.org.tw

Denmark, Finland, Norway, Iceland, Sweden

The Nordic Swan Ecolabel
The Nordic Swan Ecolabel is the official Nordic ecolabel supported by all Nordic Governments. It is among the world›s strictest and most recognised environmental certifications.

The Nordic Swan Ecolabel is a Type I environmental labelling program established in 1989 by the Nordic Council of Ministers, connect¬ing policy, people, and businesses with the mission to make it easy to make the environmentally best choice. Nordic Ecolabelling is the non-profit organisation responsible for the Nordic Swan Ecolabel.

The organisation offers independent third-party certification and support for a wide range of product areas and services, ensuring that they comply with the Nordic Swan Ecolabel's strict requirements through documentation and inspections.

30 years of experience and expertise has made the Nordic Swan Ecolabel a powerful tool that paves the way to a sustainable future by giving producers a recipe on how to develop more environmentally sustainable products, and giving consumers credible guidance by helping them identify products that are among the environmentally best.

Globally, you can find more than 25,000 Nordic Swan ecolabelled products. 93% of all Nordic consumers recognise the Nordic Swan Ecolabel as a brand, and 74% believe that the Nordic Swan Ecolabel makes it easier for them to make envi¬ronmentally friendly choices (IPSOS 2019).

Denmark, Finland, Norway, Iceland, Sweden

Securing a sustainable future

The Nordic Swan Ecolabel works to reduce the overall environmental impact from production and consumption and contributes significantly to UN Sustainable Development Goal 12: Responsible consumption and production.

To ensure maximum environmental impact, the Nordic Swan Ecolabel sets product specific requirements and evaluates the environmental impact of a product in all relevant stages of a product lifecycle - from raw materials, production, and use, to waste, re-use and recycling.

Common to all products certified with the Nordic Swan Ecolabel is that they meet strict environmental and health requirements. All requirements must be documented and are verified by Nordic Ecolabelling. Nordic Ecolabelling regularly reviews and tightens the requirements.

Therefore, certifications are time-limited and companies must re-apply to ensure sustainable development.

International website:
Nordic-ecolabel.org
National websites:
Denmark: ecolabel.dk
Sweden: svanen.se
Norway: svanemerket.no (in Norwegian)
Finland: joutsenmerkki.fi (in Finnish)
Iceland: svanurinn.is   (in Icelandic)

Thailand

The Thai Green Label Scheme was initiated by the Thailand Business Council for Sustainable Development (TBCSD) in October 1993. It was formally launched in August ١٩٩٤ by The Thailand Environment Institute (TEI) and Thai Industrial Standards Institute (TISI). The Green Label is an environmental certification logo awarded to specific products which have less detrimental impact on the environment in comparison with other products serving the same function. The Thai Green Label Scheme applies to all products and services, but not foods, beverage, and pharmaceuticals. Products or services which meet the Thai Green Label criteria may carry the Thai Green Label. Participation in the scheme is voluntary.

Thailand Environment Institute (TEI)
16/151 Muang Thong Thani, Bond Street,
Bangpood, Pakkred, Nonthaburi 11120 THAILAND
Tel. +66 2 503 3333 ext. 303, 315, 116
Fax. +66 2 504 4826-8
Website: http://www.tei.or.th/greenlabel/
Email: lunchakorn@tei.or.th

EUROPE

Established in 1992 and recognized across Europe and worldwide, the EU Eco-label is a label of environmental excellence that is awarded to products and services meeting high environmental standards throughout their life-cycle: from raw material extraction, to production, distribution and disposal. The EU Eco-label promotes the circular economy by encouraging producers to generate less waste and CO_2 during the manufacturing process. The EU Ecolabel criteria also encourages companies to develop products that are durable, easy to repair and recycle.

The EU Ecolabel criteria provide exigent guidelines for companies looking to lower their environmental impact and guarantee the efficiency of their environmental actions through third party controls. Furthermore, many companies turn to the EU Ecolabel criteria for guidance on eco-friendly best practices when developing their product lines. The EU Ecolabel helps you identify products and services that have a reduced environmental impact throughout their life cycle, from the extraction of raw material through to production, use and disposal. Recognised throughout Europe, EU Ecolabel is a voluntary label promoting environmental excellence which can be trusted.

Spain , Germany, Italy, Sweden, Greece, Portugal, Poland, Belgium, Netherlands, Estonia, Finland, Austria, Lithuania, Czech Republic, Norway, Cyprus, Ireland, Slovenia, Hungary, Romania, Croatia, Bulgaria, Malta, Slovak Republic, Latvia, Luxembourg, Iceland

Contact and more information via: http://ec.europe.eu

The evolution of clothing from its fibre stage to fabric requires a lot of processes which are harmful to our environment. So it is very important to make textile industry more sustainable. Now- a- days a wide range of techniques and innovations related to textile production have been developed to save the world from being affected by the hazardous effects of chemicals.

Sustainable fashion

Sustainable fashion, also called eco fashion, is a part of the growing design philosophy and trend of sustainability, the goal of which is to create a system which can be supported indefinitely in terms of human impact on the environment and social responsibility

100% Recycled

CHAPTER 5

Type II Environmental Labelling

Type II environmental labelling refers to the claims made on product labels in connection with business centers. This includes familiar claims such as recyclable, ozone-free, 60% phosphate-free, and the like. This type of labelling can be in the form of a mark or sentence on the product packaging. Some of them are valid environmental claims—and some can be completely misleading.

Usually, all countries have laws against deceptive advertisements, so why has the International Organization for Standardization discussed this issue? The answer is that it is not clear whether the environmental claims have a technical basis or whether the ad is meaningless.

Most countries have guidelines at the national level to help producers and consumers know what constitutes a true, scientifically valid claim.
There is a national standard on this in Canada. In Australia, the Consumer Commission has published guidance on this, and there are similar examples in other countries.

Canada

Energy Moon Ecolabel established in Coquitlam, British Columbia, Canada in 2021. ENERGY MOON is an international ecolabel focused on Energy Saving in different industries and categorized as Type II Environmental Labelling. It's defined as 'self-declared' energy saving claims made by manufacturers and businesses based on ISO 14020 series of standards, the claimant can declare the Energy Saving objectives and targets and also propose programmes for achiving the defined objectives. However, this declaration will be verifiable.

Energy Moon
Coquitlam, BC CANADA

Email: info@energymoon.org
Web: www.energymoon.org

Canada

Environmental Sustain for Future kids established in Vancouver, BC Canada in 2020. (ESFK) is an international ecolabel focused on taking care of environment for future of kids.

ESFK defined as 'self-declared' environmental claims made by manufacturers and businesses based on ISO 14020 series of standards, the claimant can declare the environmental objectives and targets in relation to taking care of environment for future kids. However, this declaration will be verifiable.

Environmental Sustain for Future Kids
Vancouver, BC CANADA

Email: info@esfk.org
Web: www.esfk.org

Eco-Textiles:

Eco-textiles refers to all fabrics, clothing and accessories that have been manufactured, produced, and processed in an environmentally-conscious manner. This manner reduces any negative impact in the form of pollution or damage to the planet.

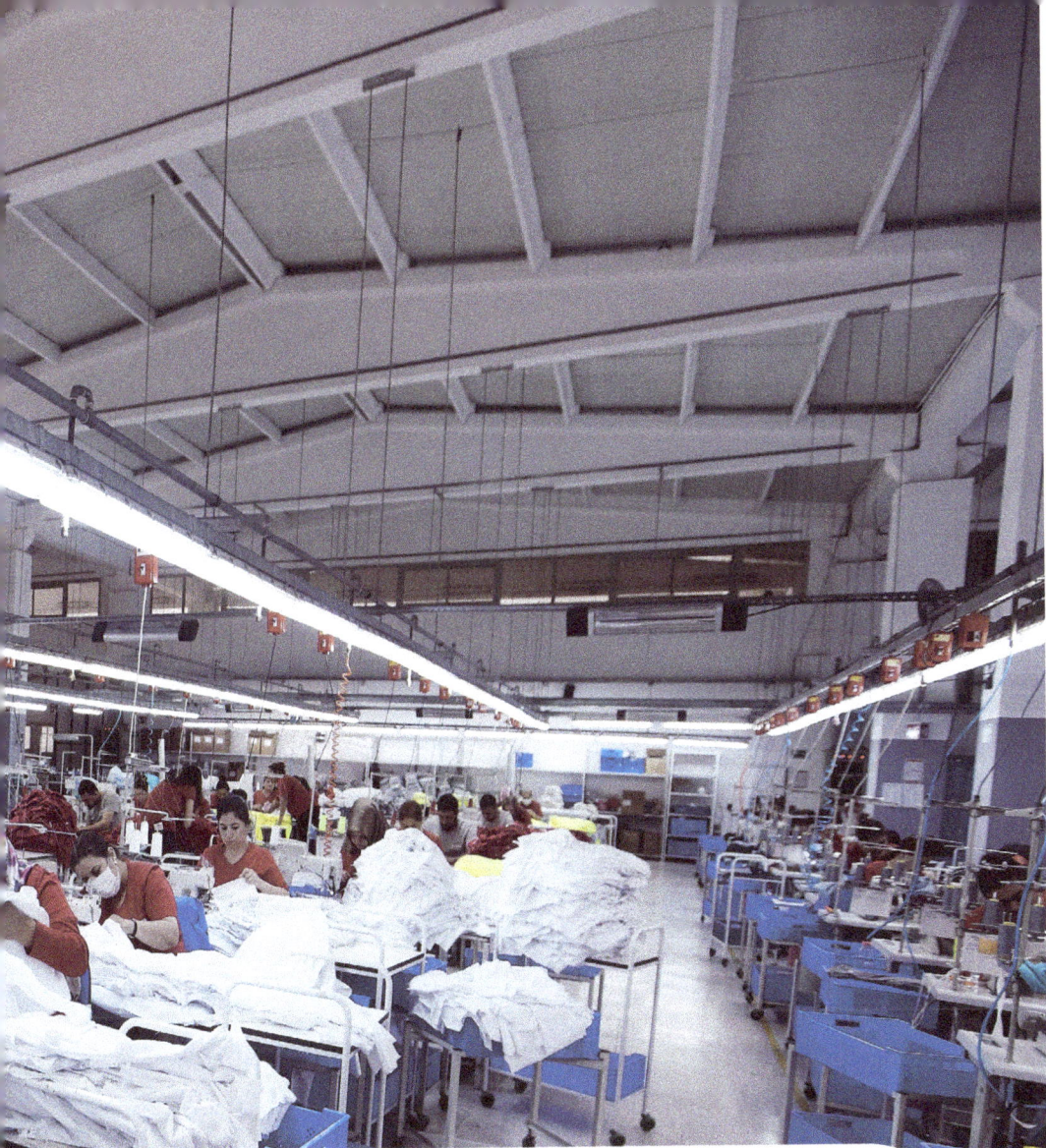

Eco-Fashion:

Eco is short for ecology, or the study of interactions between organisms and their environment. Eco-fashion is any brand or line that attempts to minimize the impact on the environment, and often the health of the consumers and the working conditions for the people that are making the clothes.

CHAPTER 6

Type III Environmental Labelling

Type III environmental labelling is a distinct form of third-party environmental labelling pattern designed to avoid the difficulties that can result from type I labelling. Technical committee for Environment of International organization for Standardization has undertaken a new project to standardize guidelines and Type III labelling methods. One of the main objections raised by industries to Type I labelling is the basis for its management.

Due to the nature of the system, less than 50% of the various products on the market can meet the criteria and qualify for Type I Labelling. As long as the industry is the main supporter of other third-party models for quality systems, it is sometimes difficult for an industry to support a program that can only benefit 15% of its members. This type of labelling is currently practiced in some countries, such as Sweden, Canada, and the United States. Choosing the right product has never been easy, but Type III labelling will help because each product can have a label that describes its environmental performance and is certified by a third-party company. Consumers can then compare labels and choose their favorite products.

CHAPTER 7

All about
'Eco-friendly' fashion and textile

ECO-FRIENDLY FASHION

ORGANIC COTTON CULTIVATION

NATURAL DYES FROM PLANTS

FABRICS FROM EASILY RENEWABLE CROPS

JEANS MADE OF REPURPOSED DENIM

The textile industry being a very good example for the most advancing and ecologically harmful industry in the world, various innovations are done in order to safeguard our mother earth. The production stages of textile include bleaching, dyeing etc...Contribute to a large extend of pollution thus making it important to make it more sustainable. Controlling pollution is as vital as making a product free from the toxic effect.So in order to safeguard our environment we must take some preventive measures and technologies that can maintain the balance of our eco system and makes the final product free from toxic effects. Generally there is really no such thing as a 100% eco friendly piece of clothing because all clothing takes water (for the fibres to grow) and energy (to make the fabric and the final garments).So, Eco-friendly clothing can be termed as a clothing made of natural fibres such as organic cotton and hemp, clothing that has been organically dyed with vegetables or any fabrics that use small amounts of water, energy and chemicals that affect the environment.

Natural fibres have intrinsic properties such as mechanical strength, low weight and healthier to the wearer that has made them particularly attractive.The word 'eco' is short for ecology. Ecology is the study of the interactions between organisms and their environment. Therefore 'eco' friendly (or 'ecology friendly') is a term to refer to goods and services considered to inflict minimal or no harm on the environment. "Think globally, act locally" is the slogan of tomorrow for the world textile industry.

ECO-FRIENDLY Textiles

Any textile product, which is produced in eco-friendly manner and processed under eco-friendly limits, is known as eco friendly textiles. It is also known as **sustainable fashion, eco fashion** and **Ecotech**. Materials can be considered as "Eco-friendly" on the basis of various factors:

- Renewability of the product
- Ecological footprint of resources - how much land it takes for the full growth of a product
- Determining the eco friendliness of a product - amount of chemicals required for the production of products.

TTAIN Eco-Review

We have created our own Fibre Eco-Review, using different resources and studies on the environmental impact of each of the fibres. Here, we have focused on the fibre production.

We have divided fibres in two main categories according to their environmental impact:

- TTAIN Recomended (TTAIN-R)
- TTAIN Not Recommended (TTAIN-NR)

RECOMMENDED

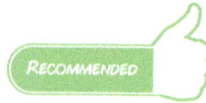

TTAIN Recomended (TTAIN-R) Fibres

Group A Plant Based	Group B Animal based	Group C Recycled	Group D Semi-Synthetic
Linen	Silk	Recycled Polyester	Pineapple, Corn, Milk, Banana, Orange
Organic Cotton	Alpaca	Recycled Nylon	Lycocell/ Tencel
Hemp	Sustainable Wool	Recycled Cotton	Algae Fibres
Ramie	Sustainable Cashmere	Recycled Wool	Cupro
Natural Rubber	Sustainable Leather	Recycled Textile Fibres	Ayurvastra
Jute	Responsible Down		Soya Bean

NOT RECOMMENDED

TTAIN Not Recomended (TTAIN-NR) Fibres

Group E Natural and Animal Based Fibres		Group F Synthetic and Semi-Synthetic Fibres	
Wool	Leather	Polyester	Acrylic
Cotton	Cashmere	Rayon	Bamboo
Down		Viscose	Vegan Leather
		Modal	Nylon
		Spandex	Polypropylene
		Aramide	PVC

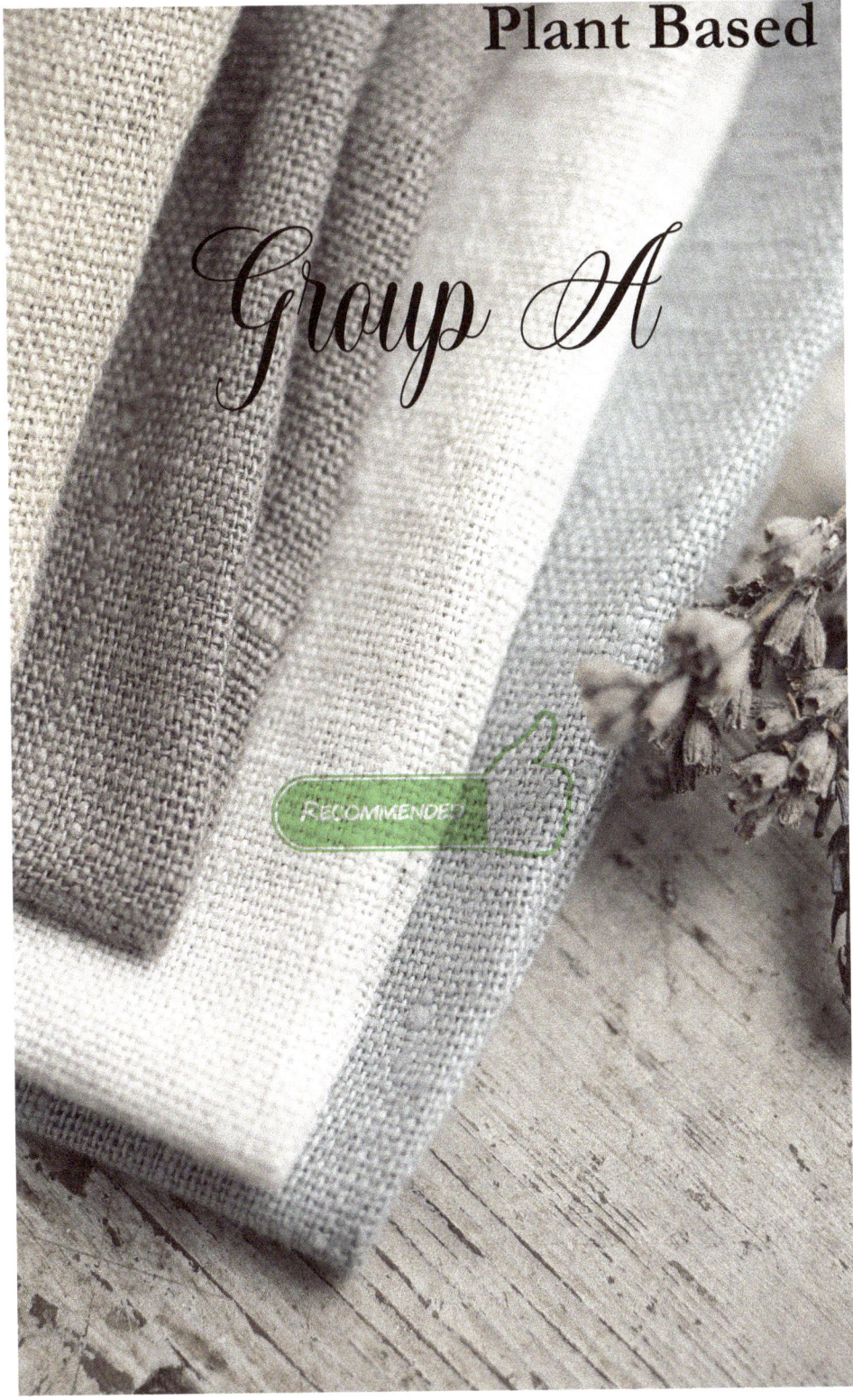

Plant Based

Group A

RECOMMENDED

Group A: Plant Based

Linen

Linen is a natural fibre which stems from the flax plant. It uses considerably fewer resources than cotton or polyester (such as water, energy, pesticides, insecticides, fertilizers).

Flax can grow in poor soil which is not used for food production. In some cases, it can even rehabilitate polluted soil. Flax plants also have a high rate of carbon absorption.

Organic Cotton

The fabric has the same quality as conventional cotton but not the negative impact on the environment. Organic cotton addresses most of the environmental challenges which conventional cotton production faces. It is grown from non-GMO seeds and without the use of pesticide, insecticide or fertilizer. Unlike conventional cotton, organic farmers use ancestral farming methods, including crop-rotation, mixed farming or no-till farming to pre-

serve the soil. Organic cotton uses up to 67% less water than conventional cotton according to some sources. Organic cotton farmers are not exposed to harmful substances. Several organizations have established certifications for organic cotton such as GOTS, (page: 29). Certification is the only proof that a product is truly organic.

Hemp

Hemp fabric comes from the plant with the same name. It is one of the fastest growing plants and it doesn't need much water, energy, pesticide, or fertilizers. The plant is very good for soil, it can be grown for many years in the same place without exhausting it. This is why hemp is considered to be eco-friendly. Hemp has very similar properties to linen. They are often difficult to differentiate. However, as hemp belongs to the same family as cannabis (although it does not have the same psychoactive effects), growing hemp is heavily regulated or prohibited in many countries.

CIRCULAR ECONOMY

MAKE · USE · REUSE · REMAKE · RECYCLE

TEXTILES

Ramie

Ramie and stinging nettle, or European nettle, are plants used to produced a fibre similar to linen. They are not very common but they are considered sustainable.

Natural Rubber

Synthetic rubber is basically plastic whereas natural rubber is made from the milk of the Hevea tree. Most of the soles of our shoes are nowadays made with synthetic rubber which is a very different thing from natural rubber. Natural rubber, therefore, comes from a renewable resource, the harvesting of rubber doesn't harm trees but actually helps the tree to flourish. It protects forests from being cut down as it gives value to the exploitation of the tree. Rubber is also easy to recycle & biodegradable. Rubber from FSC®-certified forest (Vol.1, page: 39) is even better as it ensures the good environmental management of the forest.

Jute

Jute fibre is 100% bio-degradable and recyclable and thus environmentally friendly. Jute, an edible leafy vegetable, also known as "the golden fibre", is a long, soft and shiny fibre made from the cellulose and lignin material from the jute plant. A hectare of jute plants consumes about 14.5 tonnes of carbon dioxide and releases 10.5 tonnes of oxygen. Jute also does not generate toxic gases when burnt. Jute reaches maturity quickly, between 5-7 months, making it an incredibly efficient source of renewable material, and therefore "sustainable". Jute products help in decreasing environmental pollution as its use decreases the demand for plastic bags which are non-bio degradable and pollute the surroundings. Jute bags are more useful as compared to the plastic bags as they can be used again and again. **you can simply use jute bags in lieu of non-biodegradable plastic**. Jute is also compostable by itself just like egg shells or the melon peels which means that you can sleep easy knowing that you are not contributing to the pollution or harmful clogging of our environment.

Animal Based

Group B

RECOMMENDED

Group B: Animal Based

Silk

Silk is a protein fibre spun by silkworms and is a renewable resource. Silk is also biodegradable. For these reasons, we consider silk a sustainable fibre. However, chemicals are used to produce conventional silk, so we will always consider organic silk to be a better option. Because conventional silk production kills the silkworm, animal rights advocates prefer "Peace Silk", Tussah, Ahimsa silks which allow the moth to evacuate the cocoon before it is boiled to produce silk.

Alpaca

Alpaca fibre comes from the fleece of the animal bearing the same name. Alpacas are mainly bred in the Peruvian Andes. Alpacas are much more eco-friendly than cashmere goats, because they cut the grass they eat instead of pulling it out, which allows for the grass to keep growing. Additionally, Alpacas have soft padding under their feet, which is more gentle for the soil than goat or sheep hooves.

They need very little water and food to survive and produce enough wool for 4 or 5 sweaters per year while a goat needs 4 years to produce just one cashmere sweater.

Sustainable Wool

Conventional wool is far from being as eco-friendly as we would expect. However, there are some sustainable wool options on the market which make it possible for us to dress warmly and sustainably. So far, we have found the Responsible Wool Standard (RWS), which ensures that farms use best practices to protect the land, and treat the animal decently. Certified organic wool guarantees that pesticides and parasiticides are not used on the pastureland or on the sheep themselves, and that good cultural and management practices of livestock are used. Certified organic wool is still pretty rare on the market. GOTS (page:29) seems to be the only organization certifying organic wool.

Sustainable Cashmere

As we can see in the related section, conventional cashmere has very significant consequences for the environment.

The good news is that there are a few sustainable cashmere options which address these environmental problems and give us the possibility to buy cashmere without a guilty conscience.

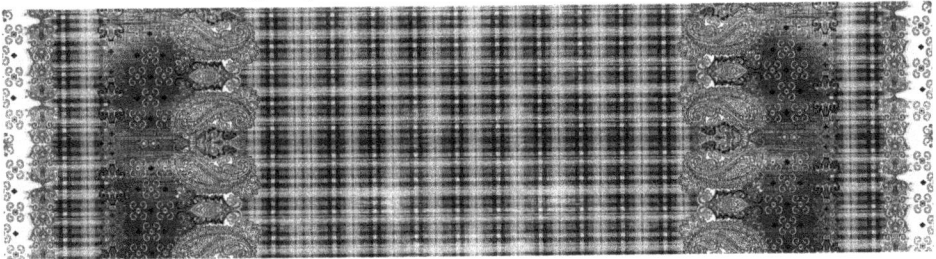

Sustainable Leather

Leather will never be an animal-friendly product: It is made of dead animal skin. However, the skin used to make leather comes from animals raised for their meat. In that sense, it uses a byproduct from another industry, so it doesn't actually need additional land and resources. Conventional leather is heavily criticized for the environmental impact of the tanning process. But leather can also be eco-friendly. There are not many options in the market yet, but they do exist.

Responsible Down

The main issue of conventional down is the live-plucking of birds which is cruel and painful to the animal. For those wanting to use down and enjoy its durability, its lightweight, and warmth, we recommend looking for certified responsible down (Responsible Down Standard) or recycled down.

Group C

RECOMMENDED

Group C: Recycled

Recycled polyester

Recycled polyester, often called rPet, is made from recycled plastic bottles. It is a great way to divert plastic from our landfills. The production of recycled polyester requires far fewer resources than that of new fibres and generates fewer CO_2 emissions.

There are 2 ways to recycle polyester: For mechanical recycling, plastic is melted to make new yarn. This process can only be done a few times before the fibre loses its quality. Chemical recycling involves breaking down the plastic molecules and reforming them into yarn. This process maintains the quality of the original fibre and allows the material to be recycled infinitely, but it is more expensive.

Although, Recycled polyester is definitely a sustainable option but, we need to be aware that it is still non-biodegradable and takes years to disappear once thrown away.

Recycled Nylon

Recycled Nylon has the same benefits as recycled polyester: It diverts waste from landfills and its production uses much fewer resources than virgin nylon (including water, energy and fossil fuel).

A large part of the recycled nylon produced comes from old fishing nets. This is a great solution to divert garbage from the ocean. It also comes from nylon carpets, tights, etc. Recycling nylon is still more expensive than new nylon, but it has many environmental advantages. A lot of research is currently being conducted to improve the quality and reduce the costs of the recycling process.

Recycled Cotton

Recycled cotton prevents additional textile waste and requires far fewer resources than conventional or organic cotton. This makes it a great sustainable option. Cotton can be recycled using old garments or textile leftovers. The quality of the cotton may be lower than of new cotton. Recycled cotton is therefore usually blended with new cotton. The production of recycled cotton is still very limited.

Recycled Wool

Recycled wool is also very sustainable option. Apart from diverting used wool garments from landfills, it saves a considerable amount of water, reduces land use for sheep grazing and avoids the use of chemicals for dyeing. Recycled wool contributes to a reduction of air, water, and soil pollution.

Recycled Textile Fibres

A lot of researches are currently going in that direction: making textile from textile waste. As we generate so much textile pre-consumer as well as post-consumer waste, it makes total sense to re-use it instead of throwing it away. However, due to the difficulty to separate fibres blend and other technological challenges, this type of textile is not yet easily available.

Semi-Synthetic

Group D

RECOMMENDED

Group D: Semi-Synthetic
Pineapple, Corn, Milk, Bannana and Orange Fibres

Pineapple Fibre (Piñatex)
Piñatex is a fibre that comes from pineapple leaves. It is considered sustainable because it uses the by-products of pineapple harvests, so there is no need for extra resources to produce it. It is used as a substitute for leather.

Corn Fibre
Corn is available in both spun and filament forms. It is derived from naturally occurring plant sugars. It balances strength and resilience with comfort, softness and drape in textiles. Corn also uses no chemical additives or surface treatments and is naturally flame retardant. Corn fibre manufacturers have claimed that these fibres can be used for sportswear, jacket, outer coat, apparels etc.

Milk Fibre
Milk Fibre was firstly introduced in 1930 in Italy & America to compete the wool. It is the new innovative Fibre and a kind of synthetic Fibre made of milk casein Fibre through bio-engineering method. It can also be used to create top-grade underwear, shirts, T-shirts, loungewear, etc. It contains seventeen amino acids & natural anti-bacterial rate is above eighty percent. Hence milk fibre has sanitarian function.

Banana Fibre
The use of banana stems as a source of fibre such as cotton and silk is becoming popular now. It is used all over the world for multiple purposes such as making tea bags or sanitary napkins to Japanese yen notes and car tyres. It is also known as musa fibre which is one of the strongest natural fibres. Banana stem, hitherto considered a complete waste, is now being made into banana-fibre cloth which comes in differing weights and thicknesses based on what part of the banana stem the fibre was taken from. The innermost sheaths are where the softest fibres are obtained, and the thicker and sturdier fibres come from the outer sheaths. High water absorbing property of this fabric makes this clothing cool to wear .

Orange Fibre

Orange Fibre is an innovative fabric made from orange skins that comes from the juice industry wastes.

Lyocell (Tencel)

Lyocell is made in a closed-loop system that recycles almost all of the chemicals used. Lyocell is a manufacturing process of rayon which is much more eco-friendly than its relatives modal and viscose. "Lyocell" is the generic name of the manufacturing process and fibre. Tencel is the brand name of the lyocell commercialized by the company Lenzing AG. Just like rayon and viscose, lyocell is more than 95% biodegradable.

Algae Fibres

Algae are being tapped as a new resource to make fibres, finishes and dyes for the textile industry. Algae bloom can provide cellulose or proteins, and in microalgae form, the species can produce non-petrochemical oils.

The food, pharmaceuticals and biofuels industries have been harnessing the ability of microalgae to produce compounds on an industrial scale for years. Now, a new generation of companies is set on putting this quality to use in developing materials and supplies for textiles, apparel and footwear. Many of the projects involving algae and textiles are still in research phase, but the landscape is changing fast.

Cupro

Cupro is an artificial cellulose fibre made from Linter Cotton (or Cotton wastes). In order to obtain the ready to weave yarn, the extracted cellulose is soaked in a bath of a chemical solution called «cuprammonium », hence the Cupro Name. All the process is made in closed-loop. The large quantities of water and chemicals used in the production of Cupro are therefore constantly reused until they are completely exhausted. The chemicals used are free of toxic or dangerous compounds for health and the environment. Cupro is also biodegradable, so it considers a good eco-friendly alternative to viscose.

Ayurvastra

Ayur vastra is a Sanskrit term made up of two words "AYUR" means "health" & "VASTRA" means "Cloth", meaning "life cloth". It is a branch of Ayurveda. Ayur vastra cloth is completely free from synthetic chemicals & toxic substances making this cloth organic, sustainable & biodegradable. Ayur vastra or medical dress is made of 100% pure organic cotton or silk, wool, jute & coir products that have been hand loomed, dyed by using various Ayurveda herbs & have medicinal qualities. Herb dyed organic fabrics act as healing agents or as an absorber through skin. Each fabric is infused with specific herbs that can help treat skin conditions. Herbs used in Ayur vastra are known to cure allergies having anti-microbial, anti-inflammatory properties Ayur vastra is extra smooth & good for transpiration that helps in recovering various diseases. It may help treat a broad range of diseases such as skin infections, diabetes, eczema, psoriasis, hypertension, high blood pressure, asthma & insomnia.

Soya Bean Fibre

Soybean fibre is a sustainable textile fibre made from renewable natural resources. The soybean protein fibre is actually made from the byproduct leftovers of soybean oil/tofu/soymilk production, which would normally be discarded.

Group E

NOT RECOMMENDED

Cotton

Cotton is mainly produced in dry and warm regions, but it needs a lot of water to grow. In some places, like India, inefficient water use means that more than 19,000 liters of water are needed to produce 1000g of cotton. In the meantime, 100 million people in India do not have access to drinking water. Although it is a natural fibre, conventional cotton is far from environmentally friendly. 97% of cotton is grown using fertilizers and genetically modified seeds. Cotton represents 11% of the pesticides and 27% of the insecticides used globally. 95% of the world's cotton farmers are located in developing countries where labor, health and safety regulations are nonexistent or not enforced most of the time. Child and forced labor are common practice. In some countries, people are forced to pick cotton for little or no pay every year.

Wool

Wool as such is a renewable natural fibre, so it could have been considered an environment-friendly option. Unfortunately, the extensive sheep farming practiced to meet the global demands has had disastrous consequences on the environment. Sheep survive by grazing, which can have a positive impact on certain types of ecosystems when it is well managed. But when the land is grazed too heavily, this leads to overgrazing. Overgrazing means that the vegetation does not have enough time to grow back before it is consumed. The soil becomes weak and vulnerable to erosion and desertification.

For example, 29% of the region of Patagonia is affected by desertification, mainly due to overgrazing by sheep which are primarily raised for their wool. Sheep also release methane, a gas that is 23 times worse for global warming than CO_2. Sheep are often subjected to insecticide baths which contain substances hazardous to the farmers. Residues of those harmful chemicals can remain in the wool and make its way into our clothes. Another concern about wool production is the poor treatment of sheep. When a sheep's fleece is removed (shearing), the shearers often hurt the animals, cutting their skin or hitting them to keep them quiet. Finally, the practice of mulesing has been widely denounced by animal rights activists. Mulesing involves removing the skin of the Merino sheep around the breech to prevent parasitic infection.

Down

Down is the layer of the fine feather of birds. Down has been used for a very long time for insulation and pillows and duvet. It is a light and warm material and very long-lasting. The main sustainability issue with down is that part of the world's supply of down feathers is directly taken ("plucked") on live birds. This practice has been largely denounced due to the suffering of the animal. It is now banned in some countries but still authorized in others. When buying down, it is essential to look for responsible down.

Leather

Leather is a controversial fibre. First of all, it is not an animal-friendly option, since it is made of dead animal skin. But environmental and social concerns related to leather are mostly linked to the tanning process: Toxic chemicals are used (chromium in 79% of cases) to transform the skins into wearable leather. Those substances are often dumped into rivers, polluting freshwater and oceans. Also, most of the tanning factory workers around the world do not wear adequate protection and suffer from skin, eye, and respiratory diseases, cancer and more due to their exposure to chemical substances.

Cashmere

Cashmere fibre comes from cashmere goat hairs. More than 77% of the world's cashmere is produced in China and Mongolia. The main environmental issue stemming from cashmere is due to the fact that goats pull the grass out by the roots when they eat instead of cutting it. As a result, the grass does not grow back, leading to land desertification. This, combined with an overpopulation of goats, results in a real environmental threat.

Mongolia is now suffering the consequences of this overgrazing through cashmere goats. The breeding of more than 21 million cashmere goats is the principal cause of the massive desertification threatening 93% of the surface of the country.

Group F

Polyester

Polyester is the most common fibre in our garment. We can find it in 55% of our clothes. Polyester is a synthetic fibre derived from petroleum, a non-renewable fossil fuel. As we know, the transformation of crude oil into petrochemicals releases toxins into the atmosphere that are dangerous for human and ecosystem health. The production of polyester also highly energy intensive. One of the major problems with this plastic fibre, is the fact that it is non-biodegradable. Furthermore, each time we wash a polyester garment, it releases 700.000 plastic microfibres, ending up in rivers and oceans and then in our food chain.

Viscose, Rayon, Modal

Viscose (also called Artificial Silk or Art Silk) is the most common type of rayon. Viscose production involves a lot of chemicals, heavily harmful to the environment when they are released in effluents. Rayon is a fibre from regenerated cellulose, generally derived from wood pulp. Rayon is usually made from eucalyptus trees, but any plant can be used (such as bamboo, soy, cotton, etc). To produce the fibre, the plant cellulose goes through a process involving a lot of chemicals, energy and water. Solvents used during the process can be very toxic to humans and to the environment. Viscose, modal, lyocell and bamboo are different types of rayon. The other substantial environmental concerns arising from rayon production is the massive deforestation involved. Thousands of hectares of rainforest are cut down each year to plant trees specifically used to make rayon. Only a very small percentage of this wood is obtained through sustainable forestry practices.

Modal, another type of rayon using beech trees with a similar process to viscose. However modal is produced by many other manufacturers who don't necessarily use sustainable processes and it is now rather easy to find sustainable fibres in the market.

Bamboo

Bamboo is usually sold as an eco-friendly textile. Which is partially true, as the bamboo plant is potentially one of the world's most sustainable resource. It grows very quickly and easily, it doesn't need pesticide or fertilizers, and it doesn't need to be replanted after harvest because it grows new sprouts from the roots. However, to turn bamboo into fibre, bamboo is processed with strong chemical solvents that are potentially harmful to the health of manufacturing workers, the consumers wearing the garment, and for the environment when chemicals are released in wastewater.

Acrylic, polyamide, nylon, polypropylene, PVC, spandex

Acrylic, polyamide, nylon, polypropylene, PVC, spandex (AKA lycra or elastane), aramide, etc, are all different types of synthetic fibres that are derived from petroleum and therefore have a very similar impact on the environment as polyester.

Vegan Leather

Vegan leather is usually made of PVC or polyurethane, which are synthetic fibres that have a similar environmental impact to polyester. It is certainly better for animal welfare, but it is not an eco-friendly option. However, some plant-based substitutes of leather exist, such as the pineapple fibre.

CHAPTER **8**

How to read Laundry Symbols

B efore reading this guide, let me ask you:

Does your head spin when you stare blankly at the laundry symbols on your most prized possessions? Have you ever bought a beautiful garment, worn it, looked at the care label symbols and thrown it in the laundry basket promising a hand wash that never happens?

We're all guilty of speed reading the label and hurriedly throwing our favorite items in the wash at 30 in the hope that it will come out of the wash just fine, right? If the responses are yes, well, this laundry symbols guide will definitely help you!

All laundry symbols have a specific meaning :

- Dry cleaning symbols
- Ironing symbols
- Drying symbols
- Hand washing symbols
- Synthetic washing symbols
- All-in-one laundry symbols

DRY CLEANING

If the care label has a small circle the manufacture is stating you must dry clean this item. If there is a little letter inside the circle it's indicating to the dry cleaner what chemical to use.

The more bars underneath the circle indicate the level of precaution the dry cleaner must take. If there is a cross over the circle symbol you should not dry clean the item. Why not leave your dry-cleaning in our expert hands and book a collection with our industry professionals.

IRONING

Your item can be ironed at any temperature if the care label iron symbol has no dots. The more dots on the iron symbol suggests the temperature of heat that can be applied:

1 dot: delicates i.e, silk and wool.
2 dots: synthetics.
3 dots: linen and cotton.

If there is a cross over the iron symbol you should not iron the item. We know how much you hate ironing, so why not get a crisp, professional press from an expert and instead spend that time doing something you love.

DRYING

If the care label has a circle inside a square, your item can be tumble dried. The more dots on the iron symbol suggests the temperature of heat that can be applied:

1 dot = low temperature
2 dots = medium temperature
3 dots = high temperature
If there is a cross over the tumble dry symbol, you should not tumble dry the item.

HAND WASHING

If the care label has a tub with a hand, your item can be hand washed or put in a delicate washing cycle of 40°C/104°F, or lower. Hand washing is better suited for delicate items, like cashmere or silk, because the wash is gentle preventing shrinking or snagging. If the care label has a twisted symbol, your item can be wrung. If the care label has a cross over the twisted symbol you should not wring the item.

SYNTHETIC WASHING

Your item can be washed in the washing machine if your care label has a tub symbol. The number on the tub symbol indicates the maximum temperature that can be applied. The more bars underneath the tub indicates a reduction of spinning and rinsing:
No bar: The item can be spun and rinsed as normal.
1 bar: Spin speed should be reduced.
2 bars: Mild wash but can be spun and rinsed as normal.
If there is a cross over the tub symbol you should not wash the item.

 Wash at or below 60°C	 Wring	Tumble Dry, No Heat	 Do Not Tumble Dry	Wet Clean	Dry Clean, Any Solvent Except Trichloroethylene, Delicate
 Wash at or below 50°C	 Do Not Wring	 Tumble Dry, High Temp	 Natural Dry	 Do Not Steam	 Dry Clean, Any Solvent, Petroleum Dry Clean, Any Solvent Only Very Delicate
 Wash at or below 40°C	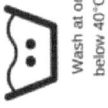 Wash at or below 95°C	 Tumble Dry, Medium Temp	 Dry Flat in Shade	 Steam	 Dry Clean, Petroleum Only Very Delicate
 Wash at or below 30°C	 Wash at or below 70°C	 Tumble Dry, Low Temp	 Dry Flat	 Do Not Iron	 Dry Clean, Petroleum Only Delicate
 Do Not Wash	 Wash at or below 60°C	 Tumble Dry, Normal	 Drip Dry in Shade	 Iron, High Temp	 Dry Clean, Petroleum Only
 Machine Wash, Delicate	 Wash at or below 50°C	 Chlorine Bleach	 Drip Dry	 Iron, Medium Temp	 Dry Clean, Any Solvent
 Machine Wash, Permanent Press	 Wash at or below 40°C	 Non-Chlorine Bleach	 Line Dry in Shade	 Iron, Low Temp	 Dry Clean
 Machine Wash, Normal	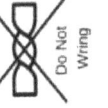 Wash at or below 30°C	 Do Not Bleach	 Shade Dry	 Iron	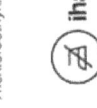 Do Not Dry Clean
 Hand Wash Normal	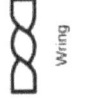 Wash at or below 95°C	 Bleach	 Line Dry	 Do Not Dry	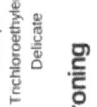 Dry Clean, Any Solvent Except Trichloroethylene, Very Delicate

Additional symbols shown:

- Iron, Low Temp
- Iron, Medium Temp
- Iron, High Temp
- Dry Clean, Reduced Moisture
- Dry Clean, No Steam
- Dry Clean, Low Heat
- Dry Clean, Short Cycle
- Do Not Wet Clean
- Wet Clean, Very Delicate
- Wet Clean, Delicate
- Dry Clean, Any Solvent Except Trichloroethylene, Very Delicate

ihateironing

Algae are being tapped as a new resource to make fibres, finishes and dyes for the textile industry. Algae bloom can provide cellulose or proteins, and in microalgae form, the species can produce non-petrochemical oils.

Port Moody, BC Canada

CHAPTER 9

Top Ten Award International Network Environmental Pioneers

Top Ten Award international Network (TTAIN) was established in 2012 to recognize outstanding individuals, groups, companies, organizations representing the best in the public works profession. TTAIN publishing books related to international Eco-labeling plans to increase public knowledge in purchasing based on the environmental impacts of products. We introduce in each volume some of the organizations that are doing their best in relation to taking care of the environmnet.

Canada

Fabdreams is a certified organic cotton bedding brand based out of Ontario, Canada. Fabdreams launched in 2019 with the wish to make your most intimate spaces of our lives warm, safe and free from chemicals. We offer pristine organic cotton beddings and bed and bath linen to our customers at an affordable price. We believe a healthy and sustainable lifestyle is not a luxury but a necessity. Through the entire cycle of creating our products, we follow sustainable, renewable and recyclable solutions right from the seed to the pod that is used for packaging. In addition to that, all our products are made in ethical, fair trade factories in India and are certified under the strict standards of Global Organic Textile Standard (GOTS), OEKO-Tex Standard 100 and Better Cotton Initiative (BCI).

Contact Detail:
Web: www.fabdreamsorganic.com
Email: interact@fabdreamsorganic.com
Phone: +1 917-475-0005

LAUNDRY SYMBOLS CHEAT SHEET

If you know these symbols...

WASH	BLEACH	TUMBLE DRY	IRON	DRY CLEAN

and these codes...

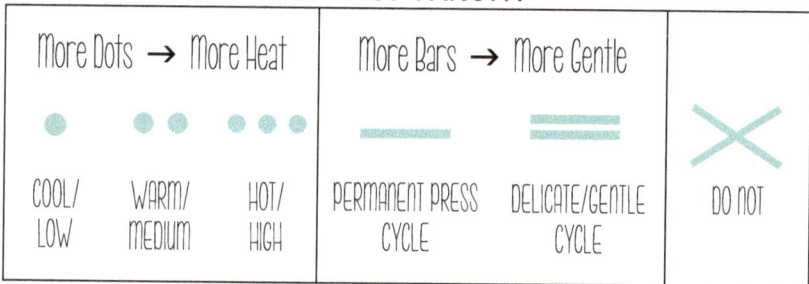

More Dots → More Heat			More Bars → More Gentle		
COOL/ LOW	WARM/ MEDIUM	HOT/ HIGH	PERMANENT PRESS CYCLE	DELICATE/GENTLE CYCLE	DO NOT

then you know what your clothes are trying to tell you!

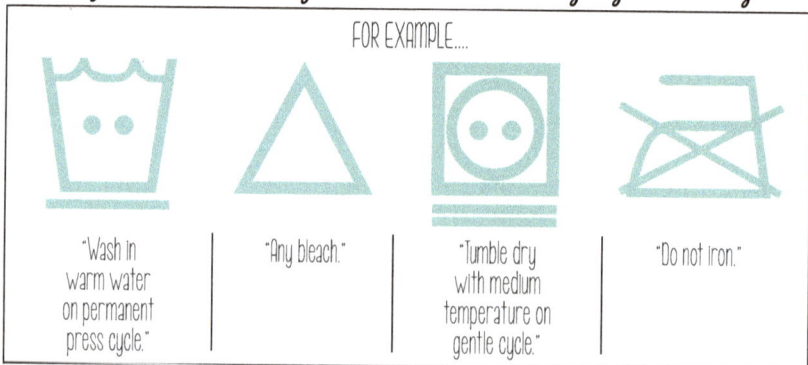

FOR EXAMPLE....

"Wash in warm water on permanent press cycle."	"Any bleach."	"Tumble dry with medium temperature on gentle cycle."	"Do not iron."

Fabric Care Symbols Guide

Machine Wash	Bleach	Tumble Dry	Dry	Iron	Dry Clean

Machine Wash

Cool/Cold

Warm

Hot

Normal

Permanent Press

Delicate/Gentle

Do Not Wash

Hand Wash

Bleach

Any Bleach
(when needed)

Only
Non-chlorine
Bleach
(when needed)

Do Not Bleach

Tumble Dry

No Heat

Low

Medium

High

Any Heat

Normal

Permanent Press

Delicate/Gentle

Do Not Tumble Dry

Dry

Line Dry/
Hang to Dry

Drip Dry

Dry Flat

Dry In The Shade

Do Not Dry

Do Not Wring

Iron

(Dry or Steam)

Low

Medium

High

No Steam

Do Not Iron

Dry Clean

Dry Clean

Do Not
Dry Clean

UN environment programme

UNEP

The United Nations Environment Programme (UNEP) is the leading global environmental authority that sets the global environmental agenda, promotes the coherent implementation of the environmental dimension of sustainable development within the United Nations system, and serves as an authoritative advocate for the global environment.

Our mission is to provide leadership and encourage partnership in caring for the environment by inspiring, informing, and enabling nations and peoples to improve their quality of life without compromising that of future generations.

Headquartered in Nairobi, Kenya, we work through our divisions as well as our regional, liaison and out-posted offices and a growing network of collaborating centres of excellence. We also host several environmental conventions, secretariats and inter-agency coordinating bodies. UN Environment is led by our Executive Director.

We categorize our work into seven broad thematic areas: climate change, disasters and conflicts, ecosystem management, environmental governance, chemicals and waste, resource efficiency, and environment under review. In all of our work, we maintain our overarching commitment to sustainability.

Website: www.uncp.org

MAKE EVERY DAY

Earth Day

- APRIL 22 -

MAKE EVERY DAY

100% ORGANIC COTTON

BEST CHOICE

Bibliography:

Andrews, R.N.L. 1998. Environmental regulation and business 'self-regulation'. Policy Sciences 31(3): 177-197.

Apodaca, Julia, "Market Potential of Organically Grown Cotton as a Niche Crop." Natural Fibres Research and Information Center, Bureau of Business Research, University of Texas at Austin, Paper presented at the Beltwide Cotton Conference in Nashville, TN, January 1992.

Asadi, J., "International Environmental Labelling, Economic Consequencies, Export Magazine, July 2001

Asadi, J. 2008. Mobile Phone as management systems tools, ISO Magazine, Vol.8, No.1

Asadi, J., Eco-Labelling Standards, National Standard Magazine, Sep. 2004.

Benitta Christy P & Dr. Kavitha S, "GO-GREEN TEXTILES FOR ENVIRONMENT", Advanced Engineering and Applied Sciences: An International Journal 2014; 4(3): 26-28

Chemical Week, 1999. Europe's Beef Ban Tests Precautionary Principle. (August 11).

CHOI, J.P. Brand Extension as Informational Leverage. Review of Eco- nomic Studies, Vol. 65 (1998), pp. 655-669.

Conway, G. 2000. Genetically modified crops: risks and promise.

Corrado, M., (1989), The Greening Consumer in Britain, MORI, London

Corrado, M., (1997), Green Behaviour – Sustainable Trends, Sustainable Lives?, MORI, london, accessed via countries. Manila, Asian Development Bank 33p.

Deo H T, "Eco friendly textile production", Indian Journal of Fibre & Textile Research Vol.26, March – June 2001,pp.61-73Dawkins, K. 1996. Eco-labeling: consumer's right-to-know or restrictive business practice? Minneapolis, Minn., Institute for Agriculture and Trade Policy.

Di Leva, C. E. 1998. International Environmental Law and Development. Georgetown Interna. Environ. Law Review 10 (2): 502-549.

Economics and Management 43, 339-359.

Eiderstroem, E. 1997. Eco-labeling: Swedish Style. Forum for Applied Research in Public Policy 141(4).

Elkington, J. and Hailes, J. 1990. The green consumer guide: You can buy products that don't cost the earth. New York, Viking Press. 96p.

EMONS, W. Credence Goods and Fraudulent Experts. RAND Journal of Economics, Vol. 28 (1997), pp. 107-119.

EMONS, W. Credence Goods Monopolists. International Journal of In- dustrial Organization, Vol. 19 (2001), pp. 375-389.

Energy.gov, Advantages and Challenges of Wind Energy, Retrieved from: https://www. energy.gov/eere/wind/advantages-and-challenges-wind-energy

Energy.gov, Advantages and Challenges of Wind Energy, Retrieved from: https://www. energy.gov/eere/wind/advantages-and-challenges-wind-energy

Environment Canada 1997. Towards Greener Government Procurement: An Environment Canada Case Study (pp. 31-46). in Greener Purchasing: Opportunities and Innovations. Environmental Protection Agency 742-R-98-009, (1998),

Environmentalist 17 (2): 125-133.

Erskine, C.C. and Collins, L. 1996. Eco-labeling in the EU: a comparative study of the pulp and paper industry in the UK and Sweden. European Environment 17 (2) : 40-47.

Erskine, C.C. and Collins, L. 1997. "Eco-labeling: Success or failure?".

Ethical Consumer, (1995), Co-op Supermarkets take up Ethics, EC36, June/July, p4

Ethical Consumer, (June 1996), Green Cons, EC41, June, p5

European Communities, Commission of the, 1996. Eco-label revision.

European Communities, Commission of the. 1996. Conservation of West Africa's forests through certification. UN Courier 157: 71-73.

European Union official website: https://ec.europa.eu/info/about-european-commission/ contact_en

Feenstra, R.C. "Exact Hedonic Price Indexes," Review of Economics and Statistics 77 (1995): 634-653.

Feenstra, R.C., and J.A. Levinsohn. "Estimating Markups and Market Conduct with Multidimensional Product Attributes," Review of Economic Studies (62 (1995): 19-52.

Forest Stewardship Council: "Principles and criteria for forest stewardship" Document 1.2: <http://www.fscoax.org>

Forsyth, K. 1999. Will consumers pay more for certified wood products? Journal of Forestry 97 (2) : 18-22.

Freeman, A. M III. The Measurement of Environmental and Resource Values. Theory and Methods. Washington D.C.: Resource for the Future, 1993.

Friends of the Earth, 1993. Timber certification and eco-labeling. London, FOE:

Geetha Margret Soundri, "Ecofriendly Antimicrobial Finishing of Textiles Using Natural Extract", Journal of International Academic Research For Multidisciplinary, ISSN: 2320 – 5083, 2014, Vol 2.

Graves, P., J.C. Murdoch, M.A. Thayer, and D. Waldman. "The Robustness of Hedonic Price Estimation: Urban Air Quality," Land Economics 64(1988): 220-233.

Halvorsen, R. and R. Palmquist. "The Interpretation of Dummy Variables in Semiloga-rithmic Equations." American Economic Review 70:474-75 (1980).

Imhoff, Dan, and Grose, Lynda, and Carra, Roberto., "Organic Cotton Exhibit," Mimeo. Simple Life and distributed the Texas Organic Cotton Marketing Cooperative, O'Don-nell, Texas (1996).

Imhoff, Dan. "Growing Pains: Organic Cotton Tests the Fibre of Growers and Manufac-turers Alike," reprinted on Simple Life's web page (simplelife.com), but first printed by Farmer to Farmer, December 1995.

Incomplete Consumer Information in Laboratory Markets. Journal of Environmental labeling.

ISO 14020, ISO 14021,ISO 14024,ISO 14025, International Organization for Standardization.

Kennedy, P.E. "Estimation with Correctly Interpreted Dummy Variables in Semilogarith-mic Equations," American Economic Review 71: 801 (1981).

Kirchho®, S., (2000), Green Business and Blue Angels.

Kraus, Jeff. Lab Technician at the North Carolina School of Textiles.

Labeling Issues, Policies and Practices Worldwide.

Lamport, L. 1998. The cast of (timber) certifiers: who are they? International J. Ecofor-estry 11(4): 118-122.

Large Scale impoverishment of Amazonian forests by logging and fire. 1999.

Lathrop, K.W. and Centner, T.J. 1998. Eco-labeling and ISO 14000: An analysis of US regulatory systems and issues concerning adoption of type II standards. Environmental

Lee, J. et al. 1996. Trade related environmental measures; sizing and comparing impacts.

Lehtonen, Markku. 1997. Criteria in Environmental Labeling: A comparative Analysis on Environmental Criteria in Selected Labeling Schemes. Geneva, UNEP. 148p.

LIEBI, T. Trusting Labels: A Matter of Numbers? Working Paper Uni versity of Bern, No. 0201 (2002).

Lindstrom, T. 1999. Forest Certification: The View from Europe's NIPFs. Journal of Forestry 97(3): 25-31. London

Losey, J.E., Rayor, L.S. & Carter, M.E. 1999. Transgenic pollen harms monarch larvae. Nature 399 20 May): p.214.

Management 22 (2) : 163-172.

Mattoo, A. and H. V. Singh, (1994), Eco-Labelling: Policy Considera-Michaels, R. G., and V. K. Smith. "Market Segmentation And Valuing Amenities With Hedonic Models: The Case Of Hazardous Waste Sites," Journal of Urban Economics, 1990 28(2), 223-242.

Mintel, (1991), The Green Consumer I, May

Mintel, (1994), The Green Consumer, Mintel Special Report

Moraga-Gonzalez, J. L. and N. Padr¶on-Fumero, (2002),

NCC, (1996a), Green Claims – a consumer investigation into marketing claims about the environment,

NCC, (1996b), Shades of Green – consumers' attitudes to green shopping, National Consumer Council,

Nelson , P."Information and Consumer Behaviour," Journal of Political Economy 78 (1970): 311-329..

Nicholson-Lord, D., (1993) 'Tis the Season to be Green, The Independent, 20 December

Nuttall, N., (1993), Shoppers can cross green products off their lists, The Times, 3 July

OCDE/GD(97)105. Paris, OECD. 81p.

OECD. "Ec-labelling: Actual Effects of Selected Programmes," OCDE/GD (97) 105, 1997, Paris. (available on line at http://www.oecd.org/env/eco/books.htm#trademono)

OECD. 1997a. Case study on eco-labeling schemes. Paris, OECD (30 Dec):

OECD. 1997b. Eco-labeling: Actual Effects of Selected Programs.

Osborne, L. "Market Structure, Hedonic Models, and the Valuation of Environmental Amenities." Unpublished Ph.D. dissertation. North Carolina State University, 1995.

Osborne, L., and V. K. Smith. "Environmental Amenities, Product Differentiation, and market Power," Mimeo, 1997.

Ozanne, L.K. and Vlosky, R.P. 1996. Wood products environmental certification: the United States perspective". Forestry Chronicle 72 (2) : 157-165.

Palmquist, R. B., F. M. Roka, and T.Vukina. "Hog Operations, Environmental Effects, and Residential Property Values," Land Economics 73(1), (1997): 114-24.

Palmquist, R.B. "Hedonic Methods," in J.B Braden and C.D. Kolstad, eds. Measuring the Demand for Environmental Improvement. Amsterdam, NL: Elsevier, 1991.

Pento, T. 1997. Implementation of Public Green Procurement Programs (22-31) in Greener Purchasing: Opportunities and Innovations. Sheffield, Greenleaf Publ. 325 p.

Perloff, J. "Industrial Organization Lecture Notes," Mimeo. University of California at Berkeley (1985).

Plant, C. and Plant, J. 1991. Green business: hope or hoax? Philadelphia, New Society Publishers 136 p.

Polak, J. and Bergholm, K. 1997. Eco-labeling and trade: a cooperative approach (Jan.): Policy in a Green Market. Environmental and Resource Economics 22, 419-

Poore, M.E.D. et al. 1989. No timber without trees. London, Earthscan. 352p.

Raff, D. M.G., and M. Trajtenberg. "Quality-Adjusted Prices for the American Automobile Industry: 1906-1940." NBER Working Paper Series, Working Paper No. 5035, February 1995.

Rastogi, J. 1998. What's Behind the Label? Complexities of Certified Wood. Ecoforestry 13 (2): 38-42.

Rinsey Antony V A, "Green and Safe Textiles", Solutions to Ecological Challenges: Multidimensional Perspectives, ISBN No: 978-81-926370-2-0, Pg 291-294, Reflection Books

Roberts, J. T. 1998. Emerging global environment standards: prospects and perils. Journal of Developing Societies 14 (1): 144-163.

Rosen, S., "Hedonic Prices and Implicit Markets: Product Differentiation in Pure Competition." Journal of Political Economy. 82: 34-55 (1974).

Ross, B. 1997. Eco-friendly procurement training course for UN HCR. : 126 p.

Ryan, S., and Skipworth, M., (1993), Consumers turn their backs on green revolution, The Times, 4 April

Salzman, J. 1997. Informing the Green Consumer: The Debate over the Use and Abuse of Environmental Labels. Journal of Industrial Ecology 1 (2): 11-22.

Sanders, W. 1997. Environmentally Preferable Purchasing: The US Experience (946-960) in Greener Purchasing: Opportunities and Innovations. Sheffield, Greenleaf Publ. 325p.

Sayre, D. 1996. Inside ISO 14000: The competitive advantage of environmental management. Delray Beach FL., St. Lucie Press. 232p.

SHAPIRO, C. Premiums for High Quality Products as Returns to Reputa- tion. Quarterly Journal of Economics, Vol. 98, No. 4 (1983), pp. 659-680.

Stillwell, M. and van Dyke, B. 1999. An activists handbook on genetically modified organisms and the WTO. Washington DC., The Consumer's Choice Council: 20 p.

Teisl, M. F., B. Roe, and R. L. Hicks. "Can Eco-labels tune a market? Evidence from dolphin-safe labeling," Presented paper at the 1997 American Agricultural Economics Association Meetings, Toronto.

THE GERSEN, C. Psychological Determinants of Paying Attention to Eco- Labels in Purchase Decisions: Model Development and Multinational Vali- dation. Journal of Consumer Policy, Vol. 23, No. 4 (2000), pp. 285-313.

Tibor, T. and Feldman, I. 1995. ISO 14000: a guide to the new environmental management standards. Burr Ridge Ill., Irwin Professional Publ. 250 p.

Torre, I. de la, & Batker, D. K. (n.d.) 1999-2000. Prawn to trade: prawn to consume. Graham WA., Industrial Shrimp Action Network (isatorre@seanet.com), [and] Asia –Pacific

Townsend, M. 1998. Making things greener: motivations and influences in the greening of manufacturing. Aldershot, England, Ashgate Publisher. 203p.

U.S. Energy Information Administration, What is U.S. Electricity Generation by Energy Source?, Retrieved From: https://www.eia.gov/tools/faqs/faq.php?id=427&t=3

U.S. Energy Information Administration, Biomass Explained, Retrieved From: https://www.eia.gov/energyexplained/?page=biomass_home

U.S. Environmental Protection Agency. National Water Quality Fact Inventory: 1990 Report to Congress. EPA 503-9-92-006, Apr. 1992.

UK Eco-labelling Board website, accessed via http://www.ecosite.co.uk/Ecolabel-UK/

US Environmental Protection Agency (EPA742-R-99-001): 40 p. <www.epa.gov/opptintr/epp>

US EPA, 1993. Determinants of effectiveness for environmental certification and labeling programs. Washington, D.C., US Environmental Protect

US EPA, 1993. Status report on the use of environmental labels worldwide. Washington, D.C., US Environmental Protection Agency (742-R-93-001 September).

US EPA, 1993. The use of life-cycle assessment in environmental labeling. Washington, D.C., US Environmental Protection Agency (742-R-93-003 September).

US EPA, 1998. Environmental labeling: issues, policies, and practices worldwide. Washington DC., Environmental Protection Agency, Pollution Prevention Division Prepared by Abt

US EPA, 1999. Comprehensive procurement guidelines (CPG) program. Washington, D.C., US Environmental Protection Agency: <www.epa.gov/cpg>

US EPA, 1999. Environmentally preferable purchasing program: Private sector pioneers: How companies are incorporating environmentally preferable purchases. Washington, D.C.,

USG, 1993. Federal acquisition, recycling, and waste prevention. Washington DC., Executive Order: (20 October).

USG, 1998. Greening the government through waste prevention, recycling, and federal acquisition. Washington, D.C., Executive Order 13101 (September).

Van der Grijp, N. 1998. The Greening of Public Procurement in the Netherlands (60-71) in Greener Purchasing: Opportunities and Innovations. Sheffield, Greenleaf Pub. 325 p.

Vanclay, J.K. 1996. Lessons from the Queensland rainforests: steps towards sustainability. J. Sustainable Forestry 3 (2/3): 1-25.

Vidal, J., (1993), Shopping for a paler shade of green, The Guardian, 7 April

Voluntary Overcompliance. Journal of Economic Behavior and Organization

Von Felbert, D. 1995. Trade, environment and aid. Paris, OECD Observer 195: 6-10.

Ward, H. 1997. Review of European Community and International Environmental Law 6 (2): 139-147.

Wasik, John, F. Green Marketing and Management: a Global Perspective, Blackwell Business: Cambridge, Mass, 1996.

West, K. 1995. Ecolabels: the industrialization of environmental standards. The Ecologist (Jan/Feb) 25: 16-20.

Worcester, R., (1995), Business and the Environment – in the aftermath of Brent Spar and BSE, MORI,

World Commission on Forests and Sustainable Development: Final Report. <http://iisd.ca/wcfsd>.

Zarrilli, S., V. Jha, and R. Vossenaar, eds. Eco-labelling and International Trade, St martin Press, Inc. New-York, 1997.

APPENDIX I: SEARCH BY LOGOS

H ere you can search the logos in this volume. It will help you to better undersand the Ecolabels you may encounter while shopping. Buying Eco-products will aid in having a better environment with minimum polution during production processes. Three important parameteres for shopping are **quality**, **price** & **environmental impacts** of the products.

Vol.3 Goto page: 29	Vol.3 Goto page: 44
Vol.3 Goto page: 38	Vol.3 Goto page: 29
Vol.3 Goto page: 44	Vol.3 Goto page: 88
Vol.3 Goto page: 33	Vol.3 Goto page: 34

Vol.3 Goto page: 43	Vol.3 Goto page: 31
Vol.3 Goto page: 44	Vol.3 Goto page: 32
Vol.3 Goto page: 57	Vol.3 Goto page: 30
Vol.3 Goto page: 35,36	Vol.3 Goto page: 37

Vol.3 Goto page: 81	Vol.3 Goto page: 81
Vol.3 Goto page: 37	Vol.3 Goto page: 43
Vol.3 Goto page: 79	Vol.3 Goto page: 79
Vol.3 Goto page: 81	Vol.3 Goto page: 79

Vol.3 Goto page: 80	Vol.3 Goto page: 80
Vol.3 Goto page: 81	Vol.3 Goto page: 81
Vol.3 Goto page: 81	Vol.3 Goto page: 81
Vol.3 Goto page: 81	Vol.3 Goto page: 81

Sustainable Textile and Fashion with Algae

Many of the projects involving algae and textiles are still in research phase, but the landscape is changing fast. The first garments treated with a wicking finish based on a microalgae-derived chemistry will be in stores in March at Tchibo, says Beyond Surface Technologies CEO, Matthias Foessel. This company, based in Muttenz, Switzerland, has been working with biotech start-up Checkerspot for the past three years to develop a textile finish using algae-generated oils. The first fruit of this partnership is a moisture management, wicking and quick drying finish for textiles, which is said to have attracted the attention of several major sports, athleisure and outdoor brands. The new finish offers the

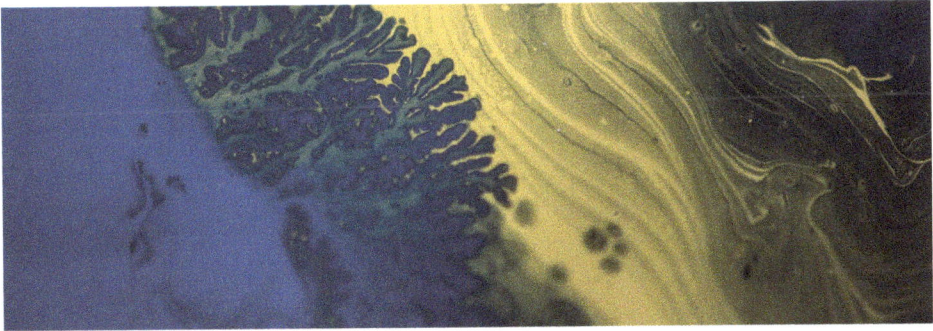

same level of performance as conventional chemicals, is a drop-in technology, and is cost-neutral compared to the company's plant seed-based chemistry, says Mr Foessel. Founded in 2008, Beyond Surface Technologies (BST) has built a portfolio of plant seed-based finishes for textiles for wicking, softening or water repellency.

Environmental Friendly Photos

Environmental friendly photos will be placed in this appendix. These photos can be received in the Top Ten Award International Network inbox from anywhere and everywhere, all over the globe. You can send your appropriate photos to us for them to be considered for publishing in one of the future, related volumes. They will be published with proper credit to the sender. The pictures can also be images of the Ecolabels existing in products within your country.

ECO-FRIENDLY FASHION

ORGANIC COTTON CULTIVATION

FABRICS FROM EASILY RENEWABLE CROPS (HEMP, BAMBOO, NETTLE)

USED RECYCLED AND STABLE UPPER AND RECYCLED WOODEN SOLES OR MADE FROM RECYCLED RUBBER RESIN

NATURAL DYES FROM PLANTS

JEANS MADE OF REPURPOSED DENIM

SUNGLASS FRAMES MADE OF RECYCLED POST CONSUMER WASTE

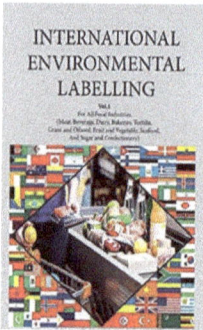

INTERNATIONAL ENVIRONMENTAL LABELLING Vol.1 For All Food Industries (Meat, Beverage, Dairy, Bakeries, Tortilla, Grain and Oilseed, Fruit and Vegetable, Seafood, And Sugar and Confectionery)	# Vol.1 Food Industries (Meat, Beverage, Dairy, Bakeries, Tortilla, Grain and Oilseed, Fruit and Vegetable, Seafood, And Sugar and Confectionery)
INTERNATIONAL ENVIRONMENTAL LABELLING Vol.2 For All Energy & Electrical Industries (Renewable Energy, Biofuels, Solar Heating & Cooling, Hydroelectric Power, Solar Power, Wind Power, Energy Conservation, Geothermal and Nuclear Power) JAHANGIR ASADI	# Vol.2 Energy & Electrical Industries (Renewable Energy, Biofuels, Solar Heating & Cooling, Hydroelectric Power, Solar Power, Wind Power, Energy Conservation, Geothermal and Nuclear Power)
INTERNATIONAL ENVIRONMENTAL LABELLING Vol.3 For All Fashion & Textile Industries (Fashion Design, The Fashion System, Fashion Retailing, Marketing and Merchandizing, Textile Design and Production, Clothing and Textile Recycling) JAHANGIR ASADI	# Vol.3 Fashion & Textile Industries (Fashion Design, The Fashion System, Fashion Retailing, Marketing and Marchandizing, Textile Design and Production, Clothing and Textile Recycling)
INTERNATIONAL ENVIRONMENTAL LABELLING Vol.4 For All Health & Beauty Industries (Fragrances, Makeup, Cosmetics, Personal Care, Sunscreen, Toothpaste, Bathing, Nailcare & Shaving, Skin Care, Foot Care, Hair Care and Other Health & Beauty Products) JAHANGIR ASADI	# Vol.4 Health & Beauty Industries (Fragrances, Makeup, Cosmetics, Personal Care, Sunscreen, Toothpaste, Bathing, Nailcare & Shaving, Skin Care, Foot Care, Hair Care and Other Health & Beauty Products)

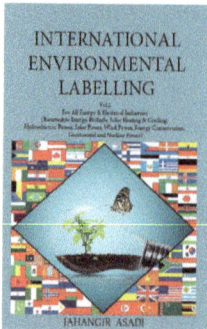

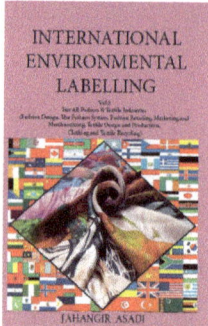

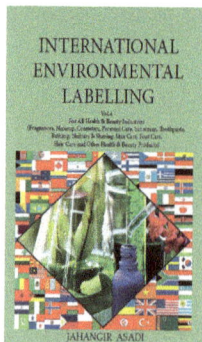

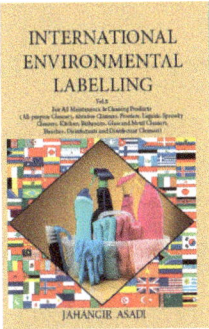

INTERNATIONAL ENVIRONMENTAL LABELLING	# Vol.5 Maintenance & Cleaning Products (All-purpose Cleaners, Abrasive Cleaners, Powders. Liquids, Specialty Cleaners, Kitchen, Bathroom, Glass and Metal Cleaners, Bleaches, Disinfectants and Disinfectant Cleaners)
INTERNATIONAL ENVIRONMENTAL LABELLING	# Vol.6 Wood & Stationery Industries (Wooden Products, Cardboard, Papers, Markers, Pens, NoteBooks. Writing Pads and Writing Sets, Pencils, White Papers, Envelopes and Organizers, Staplers and Paper Clips)
INTERNATIONAL ENVIRONMENTAL LABELLING	# Vol.7 DIY & Construction Industries (Do it yourself " ("DIY") of Building, Modifying, or Repairing, Renovation, Construction Materials, Cement, Coarse Aggregates. Clay Bricks, Power Cables, Pipes and Fittings, Plywood, Tiles, Natural Flooring)
INTERNATIONAL ENVIRONMENTAL LABELLING	# Vol.8 Agricuture & Gardening Industries (Shifting Cultivation, Nomadic Herding, Livestock Ranching, Commercial Plantations, Mixed Farming, Horticulture, Butterfly Gardens, Container Gardening, Demonstration Gardens, Organic Gardening)

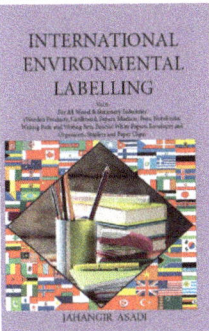

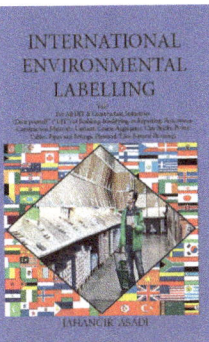

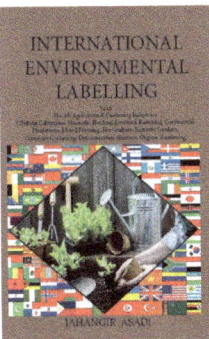

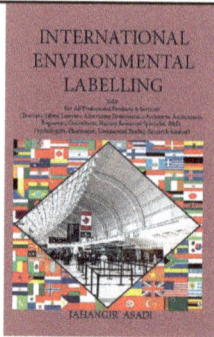

INTERNATIONAL ENVIRONMENTAL LABELLING Vol.9 JAHANGIR ASADI	# Vol.9 Professional Products & Services (Teachers, Pilots, Lawyers, Advertising Professionals, Architects, Accountants, Engineers, Consultants, Human Resources Specialist, R&D, Psychologists, Pharmacist, Commercial Banker, Research Analyst)
INTERNATIONAL ENVIRONMENTAL LABELLING Vol.10 JAHANGIR ASADI	# Vol.10 Financial Products & Services (Banking, Professional Advisory, Wealth Management, Mutual Funds, Insurance, Stock Market, Treasury/Debt Instruments, Tax/Audit Consulting, Capital Restructuring, Portfolio Management)
INTERNATIONAL ENVIRONMENTAL LABELLING Vol.11 JAHANGIR ASADI	# Vol.11 Tourism Industries (Airline Industry, Travel Agent, Car Rental, Water Transport, Coach Services, Railway, Spacecraft, Hotels, Shared Accommodation, Camping, Bed & Breakfast, Cruises, Tour Operators)
INTERNATIONAL ENVIRONMENTAL LABELLING Knowledge Test for Vol.1 to Vol.11 For all Schools and Libraries Knowledge Test JAHANGIR ASADI	## Set Box Books Vol.1-11 ## + Free Knowledge Test for Schools, Libraries, Homes and Offices all over the globe: www.TopTenAward.Net

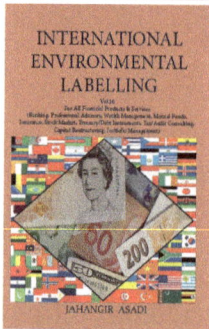

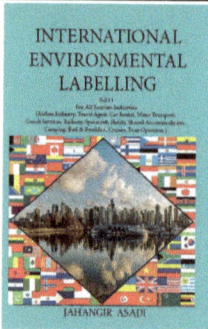

www.ingramcontent.com/pod-product-compliance
Lightning Source LLC
Chambersburg PA
CBHW040757220326
41597CB00029BB/4973